NEW THEORY OF ENERGY PRODUCTION EXPLOITING CONSTANT NATURAL FORCES, STUDY OF A GRAVITATIONAL TURBINE

"New theory of energy production exploiting constant natural forces,

Study of a gravitational turbine!

" by Sadok JABLI

Summary

Preliminary

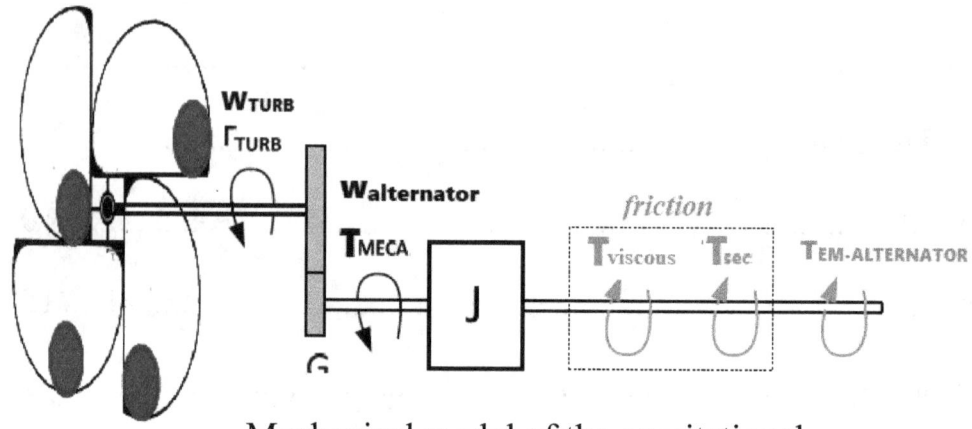

Mechanical model of the gravitational

Leonardo Da Vinci was not wrong!

Leonardo Da Vinci was not mistaken when he designed a gravitational turbine which produces mechanical energy by exploiting the gravitational force and his idea has no relation to the term "perpetual motion". I proved by precise calculations that it is a technical solution thanks to which gravity can become a source of renewable energy without violating any of the laws of thermodynamics.

I developed Leonardo da Vinci's turbine and I used it to prove the new scientific theory of energy production by means of new natural forces such as the forces of atmospheric pressure, hydrostatic pressure, gravity and the forces of certain rare metals such as neodymium and others.

I would like to invite manufacturers, engineers and researchers interested in finding a definitive solution to the current energy crisis to try to manufacture the gravitational turbine described in this research with the same dimensions and metals mentioned, and they will obtain 8MW of clean energy, this gravitational turbine is not subject to any industrial property and you can use it for free.

Our success in manufacturing these systems will bring about the biggest and fastest change in human history, as the energy will be in the hands of the people, man will rapidly transform from a failed being struggling daily for survival with the possibility of self-destruction at any moment to a self-sufficient being striving only for prosperity and peace

Leonardo Da Vinci 15th century

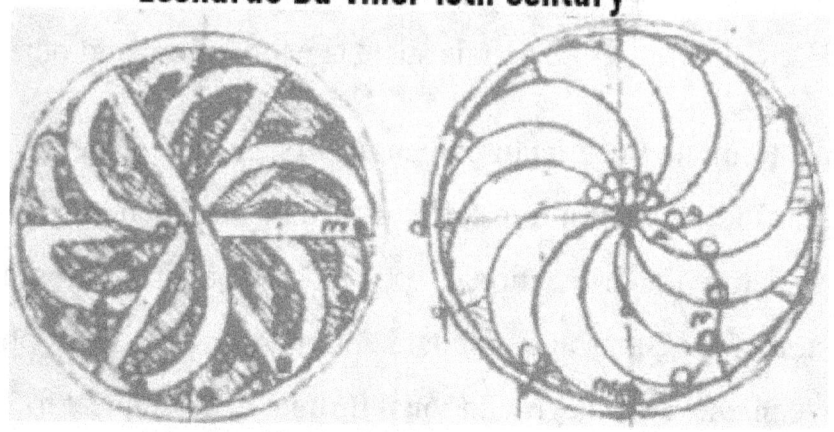

Sadok Jabli - 2021

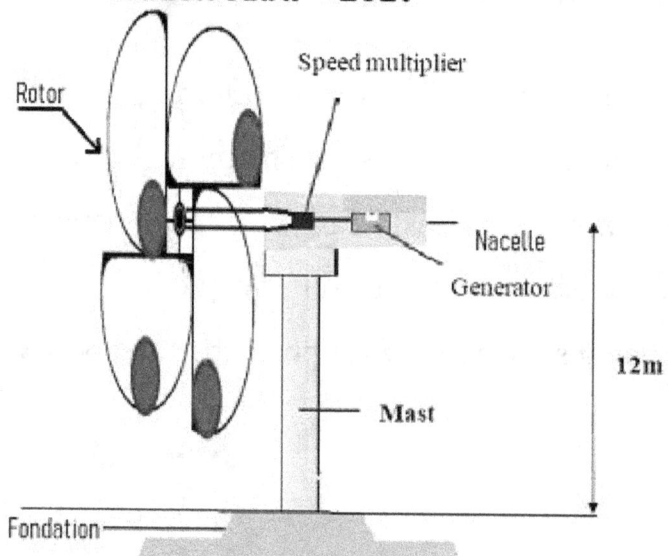

Gravitational Turbine - 8MW

6

New theory of energy production exploiting constant natural forces,

Study of a gravitational turbine!

Sadok JABLI
Independent researcher and inventor

SUMMARY

In this study I will propose a new theory of energy production exploiting constant natural forces such as: gravitational force, the force of atmospheric pressure, the force of hydrostatic pressure and even the force of permanent magnets.

I will rely in this study on a new system that I invented and which concerns the conversion of the gravitational potential energy into electric energy and relates to a new concept concerning the repetition of the work of the gravitational force provided by a gravitational turbine with magnetic bearing during a quarter revolution via a technical solution which allows to repeat the action of gravitational force and keep the dynamic equilibrium of rotation of the turbine to be able to add an alternator and convert the kinetic energy of the turbine into electrical energy.

Keywords: Alternative energy technologies, Innovation and new development, renewable energy technologies, free energy technologies

1. INTRODUCTION

In this study, I demonstrated the possibility of producing energy by stimulating the gravitational force using a technical solution which makes it move periodically and continuously.

The total work of the stimulated force during its periodic motion is the opposite of the sum of the variations of potential energy during the rotation of the "Gravitational Turbine" that I invented and which transforms this quantity of energy into electric energy

The energy produced by the stimulated force (gravitational force) is the input energy of the system.

The amount of energy is always the same regardless of the physical transformations that take place in the system.

The overall entropy of the system increases during the transformation of energy.

2. Foreword

Either the following system

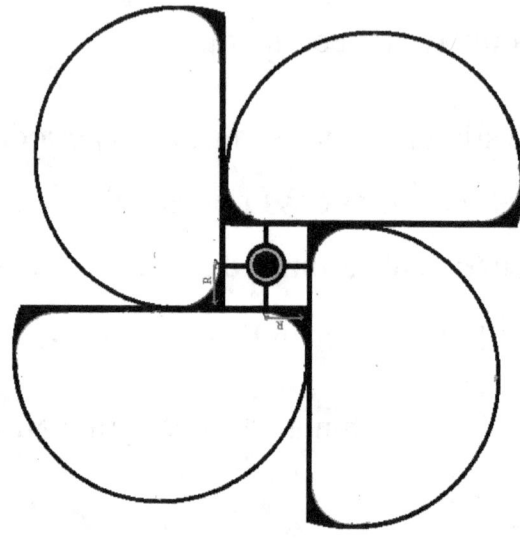

Figure.1

This system is composed of four rails in the form of semi-circles connected to each other as shown in **Figure-1** and could rotate around an axis fixed on a mast

In the state of rest, the state of static equilibrium of the turbine in relation to the ground is as shown in **Figure-2**

Point **G'** in **Figure-2** represents the center of gravity of each rail.

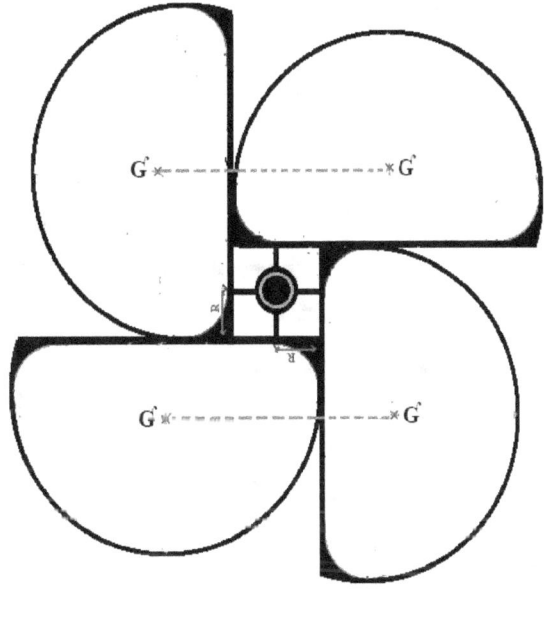

Figure 2

We add a weight **M** to the top right rail as shown in **Figure 3-1**

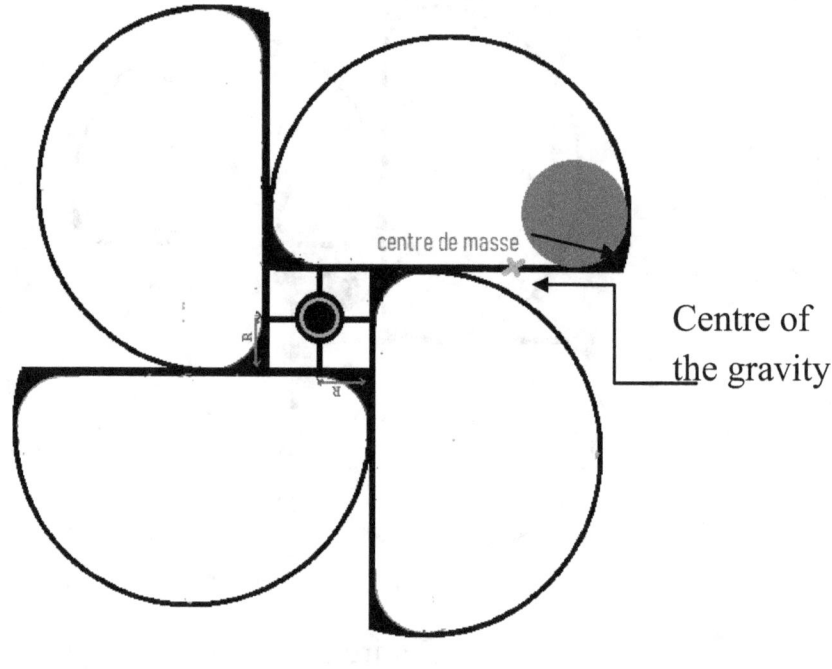

Figure.3-1

At time 1, the angular speed of the turbine (the 4 Rails + the Weight **M**) is zero and the center of mass of the turbine is at **y = 0**.

We will use a magnetic bearing to minimize friction and vibrations due to the poor distribution of masses relative to the axis of rotation.

At **y = 0** the turbine has a certain amount of potential energy.

This energy is equal to: E_p = **maximum**

On the other hand, it is motionless; it therefore does not have kinetic energy.

$E_k = 0 \text{ J}$

Arrived at its lowest point **(Figure 3-2),** the turbine no longer has potential energy

$E_p = 0 \text{ J}$

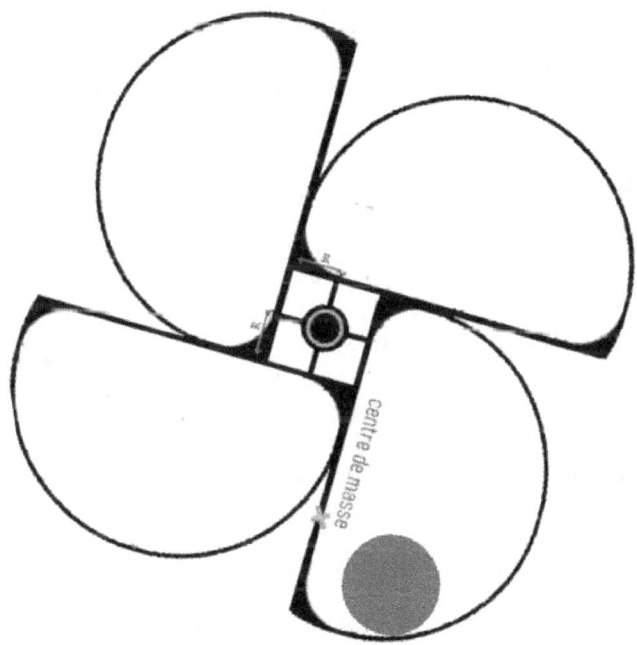

Figure.3-2

On the other hand, its kinetic energy is maximum because it reaches its maximum speed.

E_k = maximum

The amount of potential energy that the turbine has at the start of its movement is transformed into kinetic energy and part of it is transformed into thermal energy and transferred by heat to the air due to the friction existing in the mechanism

$$E_c = E_p \Leftrightarrow \frac{1}{2} J_{TURB} \, W_{TURB}^2 = M_{TURB} \, g \, \Delta y$$

The law of conservation of energy applies at all times: no matter how the turbine moves, the amount of energy is always the same.

I would like to exploit this amount of kinetic energy but the problem is that it is very limited in time and I have to provide the same amount of energy to turn the turbine and lift the weight again at point $y = 0$.

So I will propose a technical solution that allows me to repeat this transformation without violating the laws of thermodynamics and without providing energy to lift the mass **M** to its initial position.

3. Explanation

3.1 Background

The solution is to add 3 other weights of the same values in the form of cylinders capable of rolling on the rails. The turbine is now composed of 4 Rails and 4 Weights (M_1, M_2, M_3, M_4) as shown in **Figure-4**

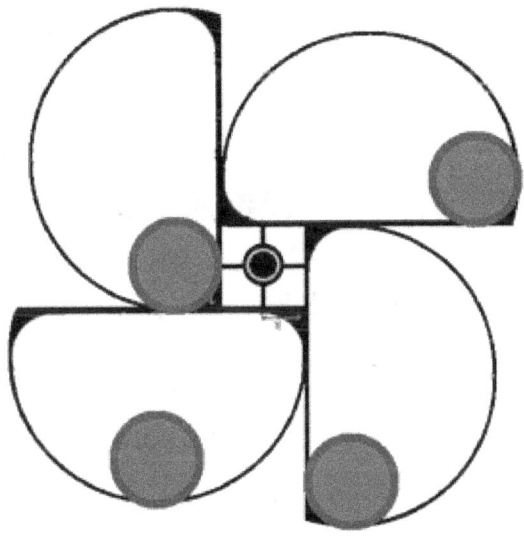

Figure.4

The 4 weights have the same value and the same shape as shown in **figure 5**:

The shape of the weights M_1, M_2, M_3, M_4

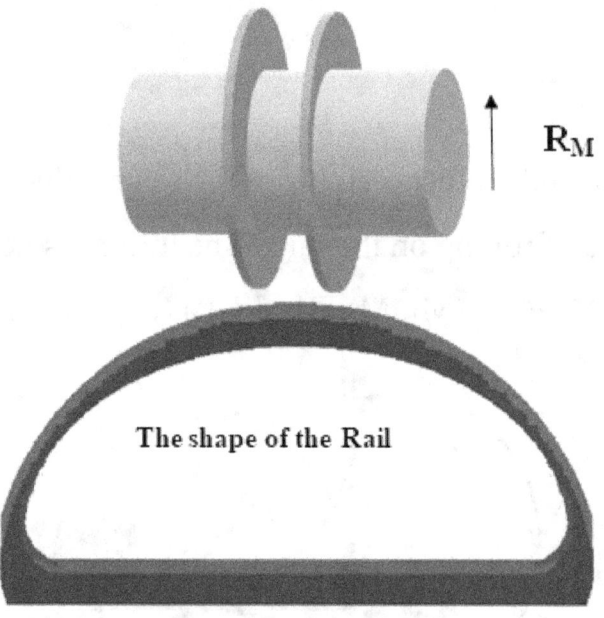

R_M

The shape of the Rail

Figure.5

The result of this solution gives a discontinuous repetitive movement of the turbine. This movement is no longer valid to be used for the production of electrical energy; I need a dynamic equilibrium of rotation to be able to add an alternator.

The following sequence of images represents the different positions of the weights on the rails during a ¼ revolution

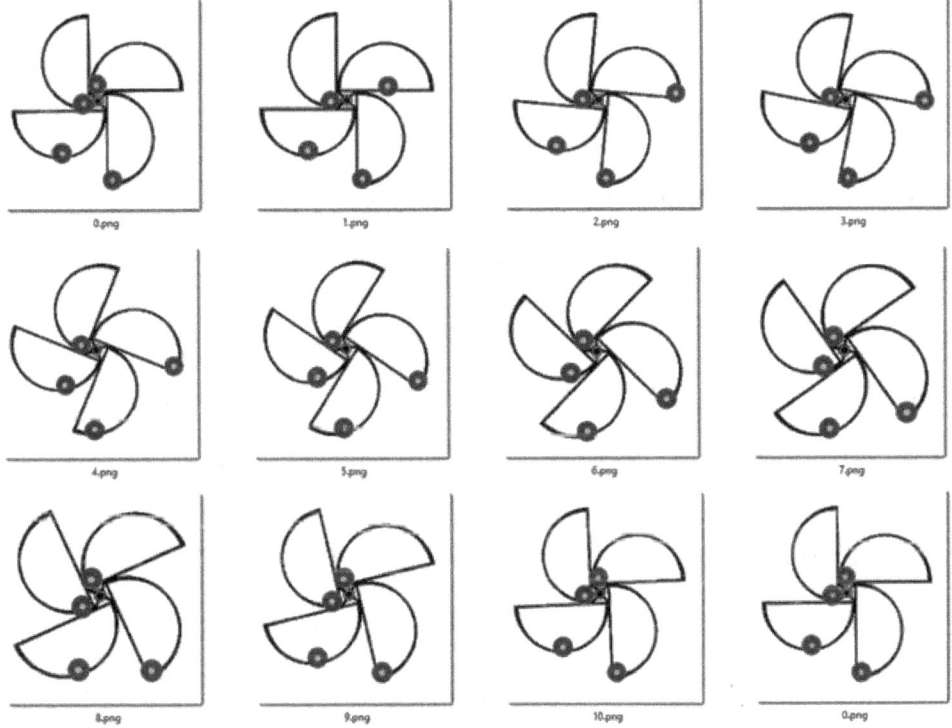

Figure-6

The discontinuous movement of the turbine poses the problem of dynamic equilibrium of rotation necessary for coupled an alternator, but before solving this problem it should be noted that it is clear that it is not a perpetual movement, such that it is defined by physicists, which designates the idea of a movement within a system, capable of lasting indefinitely without external input of energy or matter, nor irreversible transformation of the system.

Our system designates the idea of a movement limited in time which is repeated thanks to a technical solution which respects the first principle of thermodynamics which is the principle of "conservation of energy". According to this principle, the amount of energy in a system is always the same, regardless of the physical transformations that take place in that system. For our system the law of conservation of energy applies at all times during the transformation of the amount of potential energy of the turbine provides in a 1/4 revolution into kinetic energy.

Our system also respects the second law of thermodynamics which states that "any transformation of a thermodynamic system occurs with an increase in overall entropy". Entropy is the "disorder" of a system, or the degree of energy dissipation in that system. In our case some of the energy was dissipated as heat during the transformation of the turbine's potential energy quantity into kinetic energy.

I saw this movement through experience and I do not agree with the addition of the two terms "movement" and "perpetual" because if the word "perpetual" necessarily designates the violation of the laws of thermodynamics, the word "movement" necessarily designates transformation and dissipation Energy.

An object cannot move if one of these fundamental physical principles is not respected. Motion can only exist if the first and second laws of thermodynamics are respected.

Now we return to our system to solve the problem of discontinuity of the movement of the turbine and to have a dynamic equilibrium of rotation.

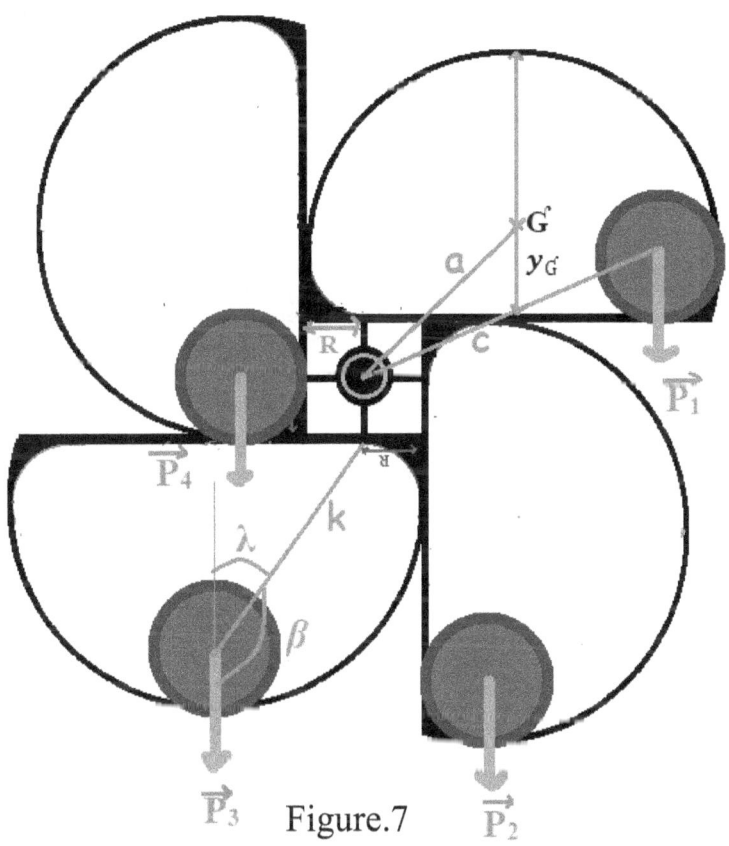

Figure.7

19

The system is made up of 4 Rails embedded to each other on which 4 weights of identical values M1, M2, M3, M4 roll as shown in **Figure-7**

Each rail is composed of 4 parts as shown in **Figure-8**:

A semicircular arc of length $\pi (l+R_M)$, a stroke of length $2l$ and two rounded ends of length $\frac{\pi R_M}{2}$ each.

$R_M = \frac{2l}{5}$ and **R=R_M** where R is the radius of the squared circle formed by the four rails and R_M radius of the rolling weight.

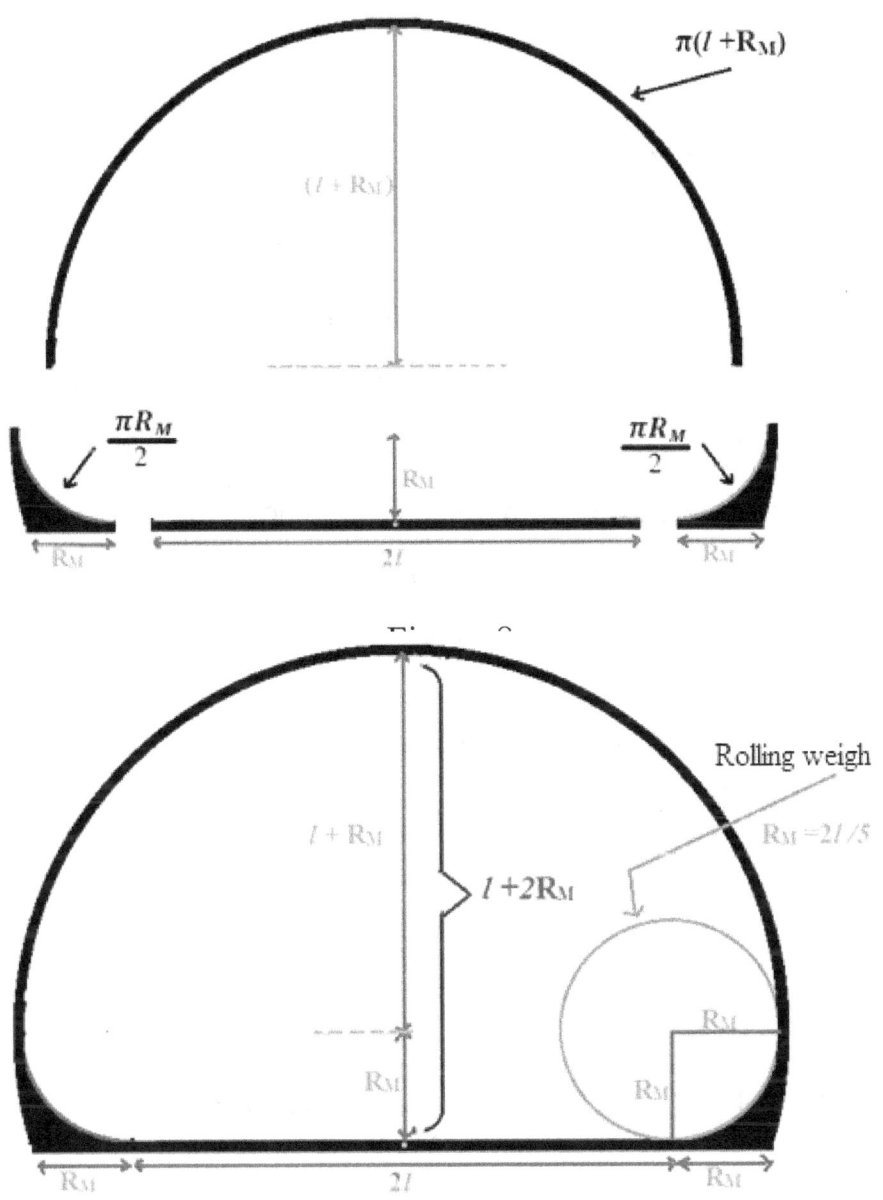

$\pi(l + R_M)$

$(l + R_M)$

$\dfrac{\pi R_M}{2}$

$\dfrac{\pi R_M}{2}$

R_M

R_M

R_M

$2l$

Rolling weigh

$R_M = 2l/5$

$l + R_M$

$l + 2R_M$

R_M

R_M

R_M

R_M

R_M

$2l$

Figure.8

The assembly (4Rails + 4Rolling weights) constitutes the turbine which rotates around an axis with a magnetic bearing to minimize friction and prevent wear of the axis.

Each weight M rolls on a rail so that the mass M_1 which rolls on the rail at the top right moves away from the axis and gives a moment of force $\Gamma_{P1} = (2l) \times (M \times g)$ for a quarter of revolution while the other masses approach the axis for the other three quarters of the revolution. This solution prevents the equilibrium between the moments of the forces of the 4 weights during one revolution (2π) of the turbine and gives a non-zero net moment applied to the turbine

Note: the position at the top right of any rail is called "the driving position", the stroke ($2l$) traversed by the weight in this position "driving stroke", the weight in this position "driving weight" and the ¼ of a revolution made by each rail in this position "the ¼ of a driving revolution"

K, a, c and R shown in **Figure-7** are distances, λ and β are angles

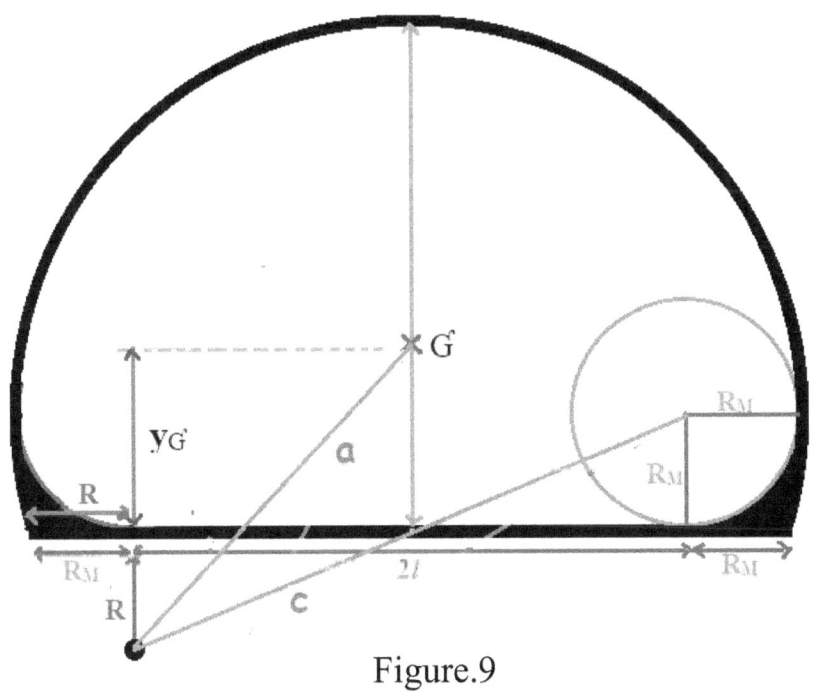

Figure.9

23

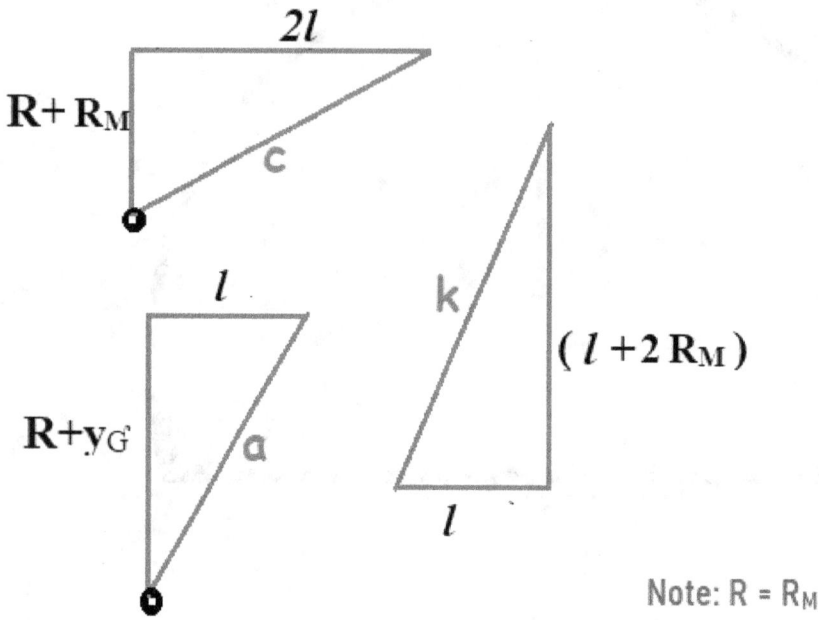

Note: R = R_M

$$a = \sqrt{(y_G + R)^2 + (l)^2}$$

$$K = \sqrt{(l + 2R_M)^2 + (l)^2}$$

$$c = \sqrt{(R + R_M)^2 + (2l)^2} = \sqrt{(2R)^2 + (2l)^2}$$

$$\cos \lambda = \frac{k^2 + (l + 2R_M)^2 - (l)^2}{2k(l + 2R_M)} \qquad \Rightarrow \lambda = x \deg \text{ (According to al-Kashi's Law}$$

of Cosines)

$\beta = 180 - \lambda \Rightarrow \sin\beta = y$

$y_G = \frac{2(l+R_M)+\pi R_M}{2\pi}$ (G is the center of gravity of the rail)

y_G is the half of the distance between G_1 (the center of gravity of the semicircular arc) and G_2 (the center of gravity of the two ends plus the stroke of the rail) (Figure-10).

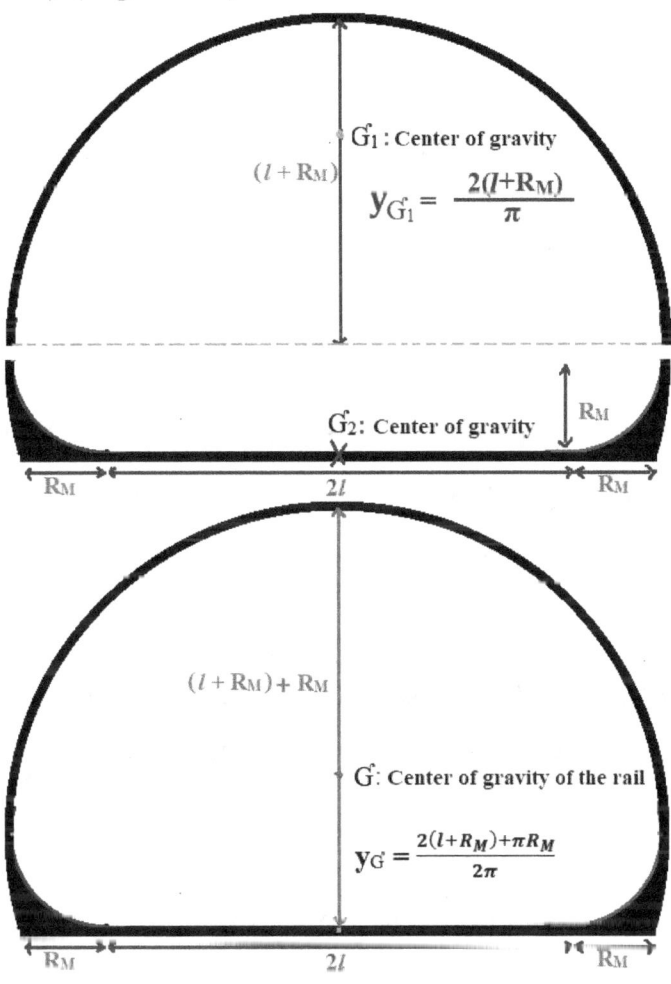

Figure.10

$$\Gamma_{P1} = (2\boldsymbol{l}) \times (M \times g)$$

$$\Gamma_{P2} = R \times P_2$$

$$\Gamma_{P3} = K \times \sin\beta \times P_3$$

$$\Gamma_{P4} = R \times P_4$$

$$\Gamma_{TURB} = (2\boldsymbol{l} \times P_1) + (R \times P_2) - (K \times \sin\beta \times P_3) - (R \times P_4)$$

$$= P_1 \times 2\boldsymbol{l} - P_3 \times K \times \sin\beta \quad \text{or } P_1 = P_3 = Mg$$

$$\Rightarrow \boldsymbol{\Gamma_{TURB}} = M \times g \times (2\boldsymbol{l} - K\sin\beta)$$

The discontinuity rotation of the turbine appears during the movement of the weight M_1 on the rail in the driving position when the average of the linear acceleration $\mathbf{a_M}$ of the weight $\mathbf{M_1}$ is less than or equal to the linear acceleration $\mathbf{a_{TURB}}$ of the turbine and when the time $\Delta\mathbf{t}$ at the end of which the turbine reaches its speed ($\mathbf{W_{TURB}}$) is greater than or equal to the duration of the time $\Delta\mathbf{t_1}$ necessary for the turbine to perform a quarter revolution.

So to avoid the discontinuity of the movement and have a dynamic equilibrium of rotation of the turbine, the following three conditions must be respected:

1st condition:

$a_{TURB} < \tilde{a}$ (a_{TURB} is strictly less than $\tilde{a}$)

$\tilde{a}$: the mean of the linear acceleration a_M of the rolling weight

a_M: linear acceleration of the driving weight during its movement on the rail for the race (**2l**)

a_{TURB}: linear acceleration of the turbine for ¼ revolution $a_{TURB} = \alpha c$

α: angular acceleration of the turbine during a ¼ revolution

c: distance between the axis of rotation of the turbine and the center of mass of the rolling cylindrical weight

The weight M has a cylindrical shape then $a_M = \frac{2}{3} g \sin\gamma$ but since γ varies between 0deg and 90deg then the linear acceleration (a_M) of the weight also increases during rolling on the driving stroke (**2l**) we will use the average of the acceleration between $\gamma = $ **1deg** and $\gamma = $ **89deg** to determine the maximum admissible value of the linear acceleration of the turbine.

$\tilde{a} = \frac{a_i + a_f}{2}$ avec $a_i = 0.11 m/s^2$ et $a_f = 6.53 m/s^2 \Rightarrow \tilde{a} = 3.32 m/s^2$

Note: the average acceleration is different from the acceleration averages', here we are talking about the acceleration averages' $\tilde{a} = \frac{a_i + a_f}{2}$

2nd condition:

$\Delta t \leq 0.7 \times \Delta t_1$

With $\Delta t = \frac{W_{TURB}}{\alpha}$ and $\Delta t_1 = \frac{\pi}{2W_{TURB}}$ $\Rightarrow$ $\frac{W_{TURB}}{\alpha} \leq 0.7 \times \frac{\pi}{2W_{TURB}}$

Δt: the time after which the turbine reaches its speed (W_{TURB}) thanks to the angular acceleration α of the turbine

$W_{TURB} = W_0 + \alpha \Delta t \Rightarrow \Delta t = \frac{W_{TURB}}{\alpha}$ with $W_0 = 0$

Δt_1: the time necessary for the turbine to make the ¼ of a driving revolution.

If the turbine makes N rpm then $60 / N = n$ seconds = the time of a single revolution and since we are looking for the time Δt_1 of ¼ of a revolution then

$\Delta t_1 = \frac{60}{N} \times \frac{1}{4} = \frac{15}{N} = \frac{15\pi}{30W_{TURB}} = \frac{\pi}{2W_{TURB}}$

3rd condition

The two ends of the rail must be rounded and of length $\frac{\pi R_M}{2}$ each, to prevent the turbine from getting stuck at the angle $\gamma = \mathbf{0deg}$ and go directly to an angle of inclination strictly greater than $\mathbf{1deg}$ (Figure.8).

The rounded end of each rail should have a length of $\frac{\pi R_M}{2}$ to avoid the shock between the rolling weight and the end of the driving rail and allow the weight to shift its speed smoothly on the rail as if it is rolling on a straight ride.

To concretely understand the system I will dimension the following mechanism entitled "Gravitational Turbine" which consists in producing electrical energy from the gravitational potential energy of the turbine transformed into kinetic energy then into mechanical energy and finally into electrical energy at the alternator level.

3.2 Sizing of a gravitational turbine of power P = 20KW

We would size a 20KW gravitational turbine to meet the electrical needs of an apartment.

Let $2l$ = 2m be the stroke of each rail and R = $2l$/5 = 0.4m the radius of the circle inscribed in the square formed by the four rails and whose center is the axis of rotation of the turbine.

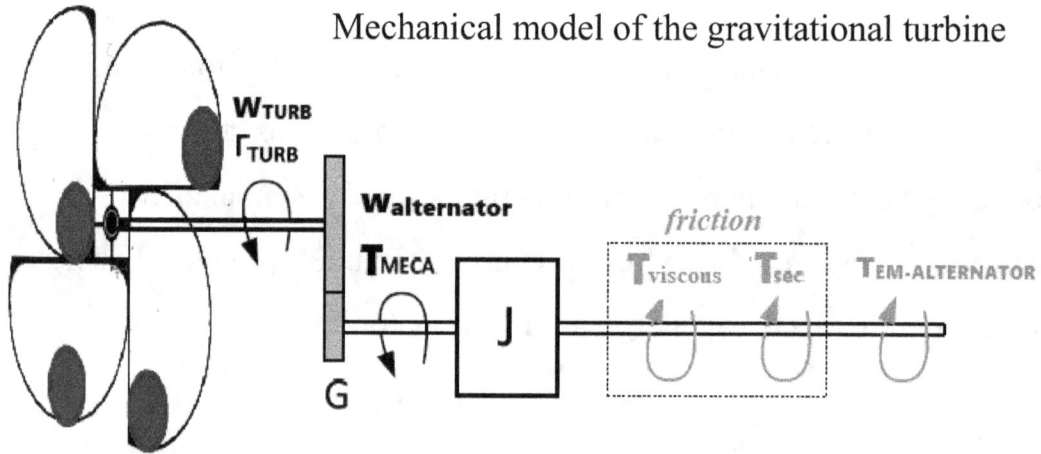

Mechanical model of the gravitational turbine

Figure.11

$$J \frac{dw_{alternator}}{dt} + fv.w_{alternator} + T_{sec} = T_{MECA} - T_{EM\text{-}ALTERNATOR}$$

With:

W_{TURB} : the rotational speed rad/s

Γ_{Turb}: Sum of the moments of the forces applied to the turbine in Nm

T_{MECA} : Turbine torque on the alternator side in Nm

$W_{alternator}$: alternator rotation speed in rad/s

G : Multiplier ratio

$T_{EM\text{-}ALTERNATOR}$: Electromagnetic torque of the alternator

J : Moment of inertia of the rotating parts referred to the alternator shaft in kg.m²

J_{TURB}: The inertia of the turbine in kg.m²

$T_{viscous}$: Viscous friction torque in N.m

T_{sec} : Dry friction torque in N.m

fv : Viscous coefficient of friction

Reminder: $T_{viscous} = fv.w_{alternator}$

$W_{alternator} = G \; W_{TURB}$

We express the mechanical torque $\mathbf{T_{MECA}}$ applied to the shaft of the alternator by the turbine:

$$T_{MECA} = \frac{P_{TURB}}{W_{alternator}}$$

Let $\quad \mathbf{W_{aternator}} = 3000$ rpm $= 314.15$ rad/s

$\quad P_{TURB} = 20KW\rangle$

$\mathbf{T_{MECA}} = \dfrac{20000}{314.15} = 63.66$ Nm

I would like the turbine to rotate at 8 rpm

$W_{TURB} = 0.837$ rad/s $\Rightarrow G = \dfrac{W_{alternator}}{W_{TURB}} = 375$

$\Gamma_{TURB} = 375 \; . \; T_{MECA}$

$\quad = 23873.24$ Nm

$\quad l = 1m$

$\quad R_M = 2/5 = 0.4m$

$R = R_M = 0.4m$

$(l + R_M) = 1.4m$

$(l + 2R_M) = 1.8m$

$(R + R_M) = 0.8m,$

$2l + R_M = 2.4m$

$y_G = \frac{2(l+R_M)+\pi R_M}{2\pi} = 0.64m$

$a = \sqrt{(y_G + R)^2 + (l)^2} = 1.44m$

$K = \sqrt{(l + 2R_M)^2 + (l)^2} = 2.05m$

$c = \sqrt{(2R)^2 + (2l)^2} = \mathbf{2.15m}$

$\cos \lambda = \frac{k^2+(l+2R_M)^2-(l)^2}{2k(l+2R_M)} = 0.87 \Rightarrow \lambda = 29.19deg$ (According to al-Kashi's Law of Cosines)

$\beta = 180 - 29.19 = 150.8deg \Rightarrow \mathbf{sin\beta = 0.48}$

$\Delta t_1 = \frac{\pi}{2W_{TURB}} = 1.87\ s$

$\Delta t = \dfrac{W_{TURB}}{\alpha}$ et $\alpha = \dfrac{\Gamma_{TURB}}{J_{TURB}}$ We need to calculate J_{TURB} to know the angular acceleration of the turbine

The inertia of the turbine is composed of the inertia of the 4 rails plus the inertia of the 4 rolling weights

Inertia of the 4 Rails

Inertia of the 4 Rails= 4 J_{Rail} = 4 $m_{Rail}\, a^2$

Inertia of the 4 rolling weights

The path followed by the 4 weights M_1, M_2, M_3, M_4 around the axis is not circular and has four parts with different speeds which leaves the calculation of the total inertia of the rolling weights very complicated.

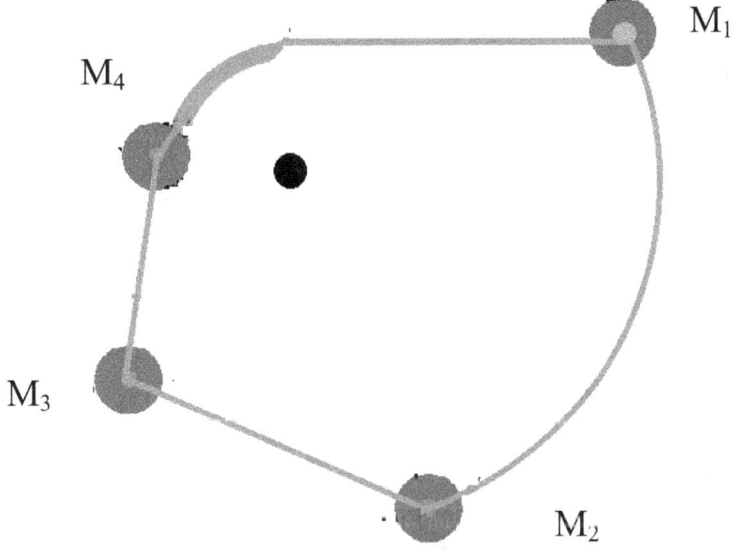

Figure.12

But according to my calculations the closest value to the total inertia of these rolling weights during the rotation of the turbine can be given by the following formula:

$$J_{\text{tot-rolling-weights}} = 2\,J_{M1} + 2\,J_{M4} = 2M_1\,c^2 + 2M_4\,R^2 = 2M(c^2 + R^2)$$

So $J_{\text{TURB}} = 4\,J_{\text{Rail}} + 2\,J_{M1} + 2\,J_{M4}$

$$-\,4\,m_{\text{Rail}}\,a^2 + 2M_1\,c^2 + 2M_4\,R^2 = 4\,m_{\text{Rail}}\,a^2 + 2M(c^2 + R^2)$$

35

Finding the mass of the rolling cylindrical weight

$M_1=M_2=M_3=M_4=M$

$\mathbf{\Gamma_{TURB}} = Mg((2l - K\sin\beta)$

$\Rightarrow M = \dfrac{\Gamma_{TURB}}{g(2l-K\mathbf{sin\beta})} = 2395.23$ kg

$\mathbf{M= 2395.23\ kg}$ with $M=M_1=M_2=M_3=M_4$

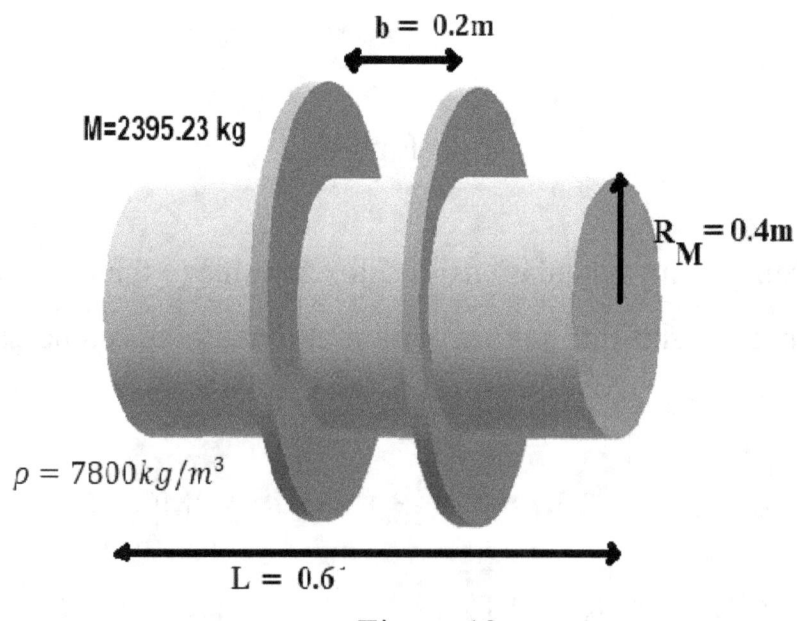

Figure.13

Find the mass of the rail

To ensure that the rails can withstand the rolling masses, we have chosen to use a high resistance material and we determine the dimensions of the rails in order to allow them to withstand the forces that require them without damage.

Table of tensile yield strength of common materials

Material	Nuance	Re (MPa)
Hardening Low Alloy Stee	30 Cr Ni Mo 16 (30 CND 8)	700 to 1 450
Fiberglass	"R", high performance	3200
Carbon fiber	Young's high modulus "HM"	2500
Carbon fiber	High strength "HR"	3200

Condition of resistance:

The resistance materials study verifies that the effective stress σ to which the part is subjected remains less than the maximum allowable stress σ_{maxi}. This is the condition of resistance.

$$\sigma_{maxi} \leqslant Rpe = Re/s \text{ ou encore } \sigma = \frac{Mfz}{\frac{IGz}{v}} \leq Rpe$$

$\frac{IGz}{v}$: Flexural modulus et $IGz = \frac{bh^3}{12}$

v: distance between the mean plane and the furthest fiber

Re: elastic limit

S: safety factor

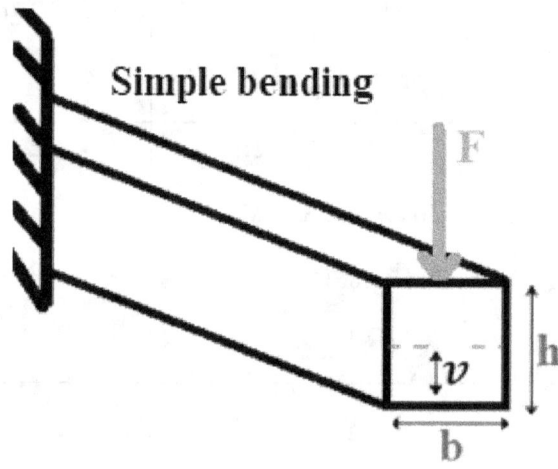

Figure.14

Allowable stress value

We can adopt, in the absence of any specific regulations, the following permissible constraint in the case of preliminary draft:

Ratings:

Re: elastic limit

S: safety factor

Sfat: safety coefficient with regard to fatigue

Kc: shock coefficient

In our case we will study the longest part of the rail since it is composed of four parts:

A semi-circular arc of length $\pi\,(l+R_M)$, a stroke of length $2l$ and two rounded ends of length $\frac{\pi R_M}{2}$ each,

With $\quad 2l = 2$ m , $\pi\,(l + R_M) = 4.398$ m and $\frac{\pi R_M}{2} = 0.62$m

We will choose carbon fiber:
$$\begin{cases} \text{Re} = 3200\text{Mpa} \\ \rho = 1800kg/m^3 \end{cases}$$

This will allow us to know σ_{maxi} and know the distance **h** to correctly size the rail.

Shear force and bending moment diagram

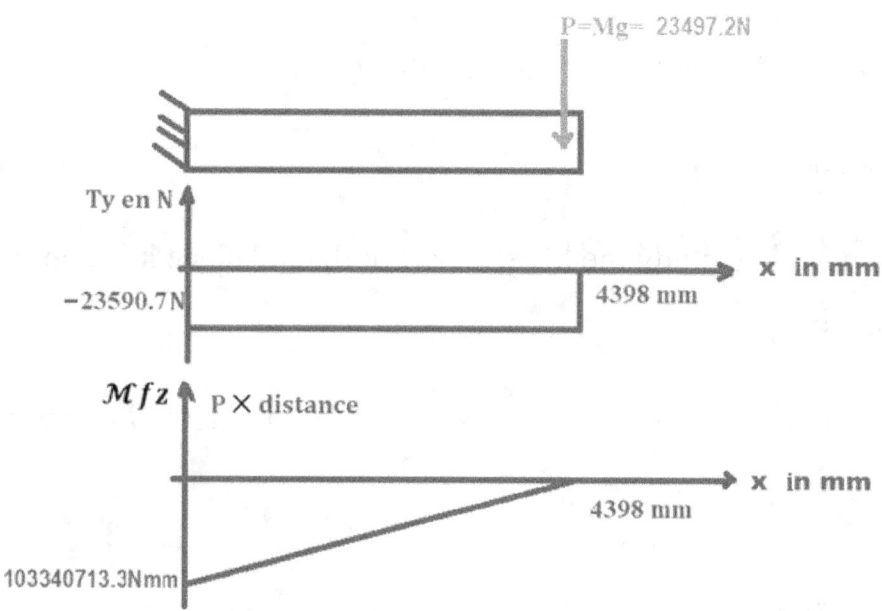

Maximum bending stress

$$\sigma_{\text{maxi}} = \frac{Re}{S \times S_{fat} \times k_c} = \frac{3200}{1.5 \times 1.5 \times 2} = \frac{3200}{4.5} = 711.1 Mpa$$

With Re =3200Mpa : elastic limit

S =1.5 $\Rightarrow$ safety factor

S_{fat} = 1.5 $\Rightarrow$ Fatigue Safety Factor

K_c = 2 $\Rightarrow$ Shock factor

We have

$$\begin{cases} * \; \sigma_{maxi} = \dfrac{Mfz}{\dfrac{IGz}{v}} \\ * \; IGz = \dfrac{bh^3}{12} \\ * \; v = \dfrac{h}{2} \end{cases}$$

$\Rightarrow \sigma_{\text{maxi}} = \dfrac{Mfz}{\dfrac{\frac{bh^3}{12}}{v}}$

This equality contains two unknowns' **b** and **h**

We chose **b = 0.2m = 200mm** and we search **h=?**

$$\frac{IGz}{v} = \frac{\frac{bh^3}{12}}{v} = \frac{bh^3}{6h} = \frac{bh^2}{6} \Rightarrow \sigma_{maxi} = \frac{Mfz}{\frac{bh^2}{6}} \Rightarrow h^2 = \frac{6Mfz}{b\,\sigma_{maxi}}$$

We choose the longest part $\pi\,(l + R_M)$ of the rail to calculate Mfz

$$Mfz = Mg\,\pi\,(l + R_M) = 103340713.3 Nmm$$

$$\sigma_{maxi} = 711.1 Mpa$$

$$h^2 = 4359.7 \Rightarrow h = 66mm = 0.066m$$

h must be greater than 6.6cm so that the effective stress σ to which the rail is subjected remains less than the maximum admissible stress σ_{maxi}

We choose $h = 10cm = 0.1$ m

The mass of each rail is therefore:

$$m_{RAIL} = bh\rho\pi\,(l + R_M) + \rho bh2l + \rho bh\pi\, R_M$$

$$= \rho bh(2l + \pi\,(l + R_M) + \pi\, R_M)$$

$$= \rho bh(2l + \pi\, l + 2\,\pi\, R_M)$$

$$= 275.57kg$$

Now that we know the masses of the 4 rails and the 4 rolling weights we can calculate the total inertia of the turbine to determine its angular acceleration.

$$J_{TURB} = 4\ J_{Rail} + 2\ J_{M1} + 2\ J_{M4}$$

$$= 4\ m_{Rail}\ a^2 + 2M_1\ c^2 + 2M_4\ R^2$$

$$\mathbf{= 4\ m_{Rail}\ a^2 + 2M(c^2+R^2)}$$

With $a^2 = (y_G + R)^2 + (l)^2 = 2.08$

$c^2 = 4l^2 + 4R^2 = 4.64$

$J_{TURB} = 4\times(275.57)\times(2.08) + 2\times(2395.23)(4.64 + 0.16)$

$J_{TURB} = 25286.95$ kgm²

$$\alpha = \frac{\Gamma_{TURB}}{J_{TURB}} = \frac{23873.24}{25286.95} = 0.94 \text{ rad/s}^2$$

$a_{TURB} = c\ \alpha = 2.03\text{m/s}^2 < \tilde{a} = 3.32\text{m/s}^2$

(a: the mean of the linear acceleration a_M of the driving rolling weight)

a_{TURB} : linear acceleration of the turbine of radius **c** during the ¼ of a driving revolution

(a_{TURB} is strictly less than **ã**)

$$\Delta t = \frac{W_{TURB}}{\alpha} = \frac{0.83}{0.94} = 0.78s < 0.7\Delta t_1 = 1.03s$$

So $\Delta t < 0.7 \times \Delta t_1 \Rightarrow$ the turbine reaches the speed of rotation (**W_{TURB}**) before the end of a driving quarter revolution

and **a_{TURB}** < **ã** $\Rightarrow$ This allows the turbine to function as a flywheel during the stroke (2*l*) of the driving weight so the rotational discontinuity disappears after a few revolutions and the turbine finds its dynamic rotational equilibrium at an angular speed very close to the theoretical speed (W_{TURB}=0.83rad/s).

Note: if the acceleration averages' of the driving weight is higher than the linear acceleration of the turbine, the actual angular speed will be closer to the theoretical angular speed (**W_{TURB}**)

The two necessary conditions are met for this example and the turbine will find its dynamic rotation equilibrium after a few turns but a vibration problem could appear during rotation because of the variations in the speed of the weights on the rails if the inertia of the four rails is too low compared to the inertia of the driving weight most affected by the sudden drop in speed.

To solve this problem I will increase the inertia of the rails without touching 2nd condition of dynamic equilibrium $\Delta t \leq 0.7\Delta t_1$

We must not exceed 70% of Δt_1 when we want to decrease the acceleration of the turbine by increasing its inertia.

The value of Δt max-admissible is therefore $\Delta t_{\text{max-admissible}} = 0.7\Delta t_1$

So for our example above we have:

$\Delta t_{\text{max-admissible}} = 0.7 \times 1.87 = 1.3\text{s}$

$\Rightarrow \alpha_{\text{min-admissible}} = W_{\text{TURB}} / \Delta t_{\text{max-admissible}} = 0.63 \text{ rad/s}^2$

Or we have $\alpha_{min-admissible} = \dfrac{\Gamma_{TURB}}{J_{MAX-TURB}}$

$\Rightarrow J_{MAX-TURB} = \dfrac{\Gamma_{TURB}}{\alpha_{min-admissible}} = \dfrac{23873.24}{0.63} = 37894 \text{ kgm}^2$

With $J_{MAX-TURB}$: maximum admissible inertia of the turbine

We will therefore add $J_A = 12607\text{kgm2}$

$$J_{\text{MAX-TURB}} = J_{\text{TURB}} + J_A \quad \text{with } J_A\text{: added inertia}$$

This mass can be harnessed to secure the system by adding a top rail inside each turbine rail to secure the rolling weights in the event of an earthquake.

$$J_A = 4 \times J_{\text{new-rail}} \quad \Rightarrow \quad J_{\text{new-rail}} = \frac{J_A}{4} = 3151.77 \text{ kgm}^2$$

$$J_{\text{new-rail}} = m_{\text{new-rail}} \times a^2$$

$$\Rightarrow \quad m_{\text{new-rail}} = \frac{J_{new-rail}}{\alpha^2_{min-admiss}} = \frac{3151.77}{0.39} = 7940.96 \text{kg}$$

The mass of each inner safety rail is therefore:

$m_{\text{new-rail}} = 7941\text{kg}$ The material used to make this inner rail is steel

This same turbine could produce up to 150KW if we increase the value of each rolling weight up to 18 tons taking into account the resistance of each rail which must have a value of h = 19cm for this new power.

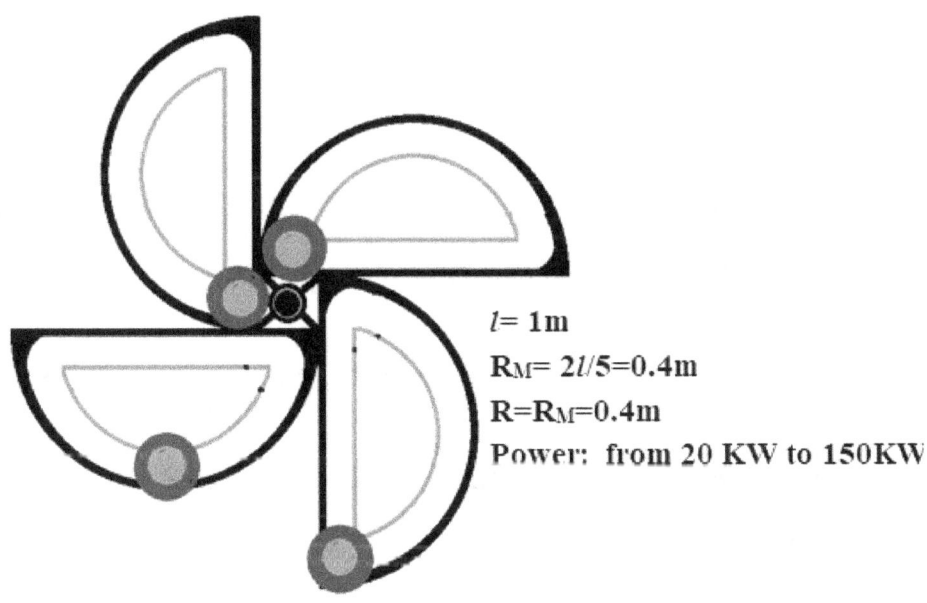

$l= 1m$
$R_M= 2l/5=0.4m$
$R=R_M=0.4m$
Power: from 20 KW to 150KW

Figure.15

3.3 Sizing of a gravitational turbine of power P = 8MW

If we would like to size a turbine with a power of **8MW** with a stroke of $2l = 8$m for a speed of 5rpm we will have the following result:

l =4m, R_M =1.6, R =1.6m, c =8.6m, k =8.23m, y_G =2.58m, a =4.49m

Mass of the cylindrical rolling weight: M = 384 601 kg of steel and length L = 6.13 m

Rail mass: 44092kg of carbon fiber with b = 1m and h = 0.8m

Rotor diameter: 19.46m

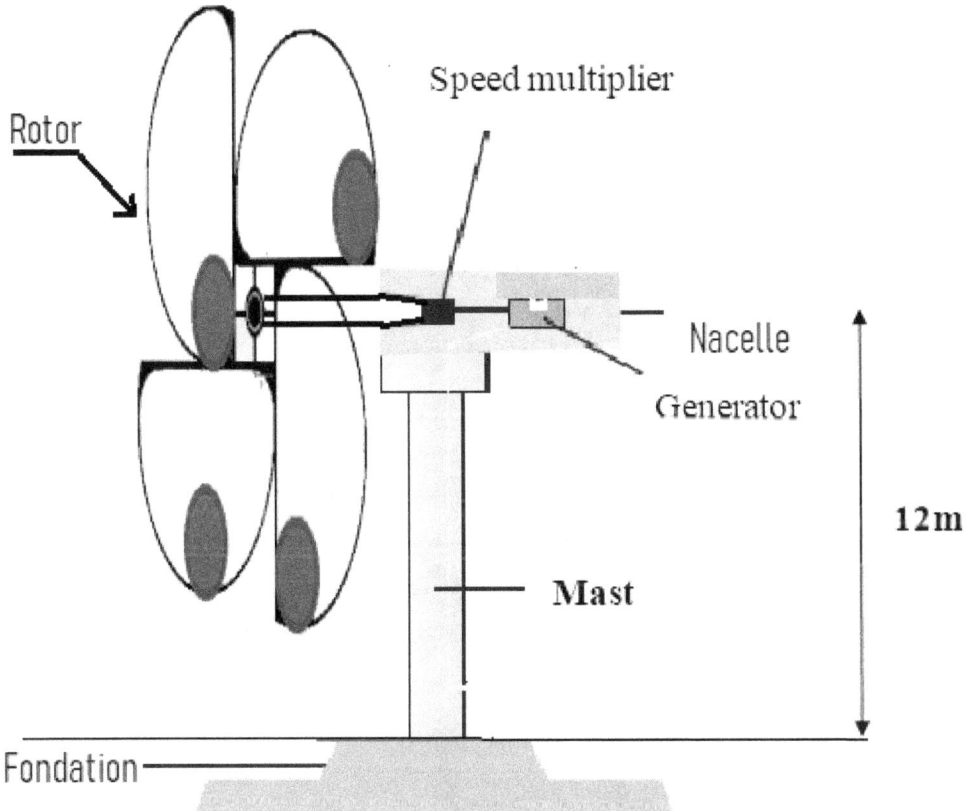

Rotor

Speed multiplier

Nacelle

Generator

12m

Mast

Fondation

Gravitational Turbine - 8MW

Figure.16

Comparison between a wind turbine and a gravitational turbine for a power of 8MW

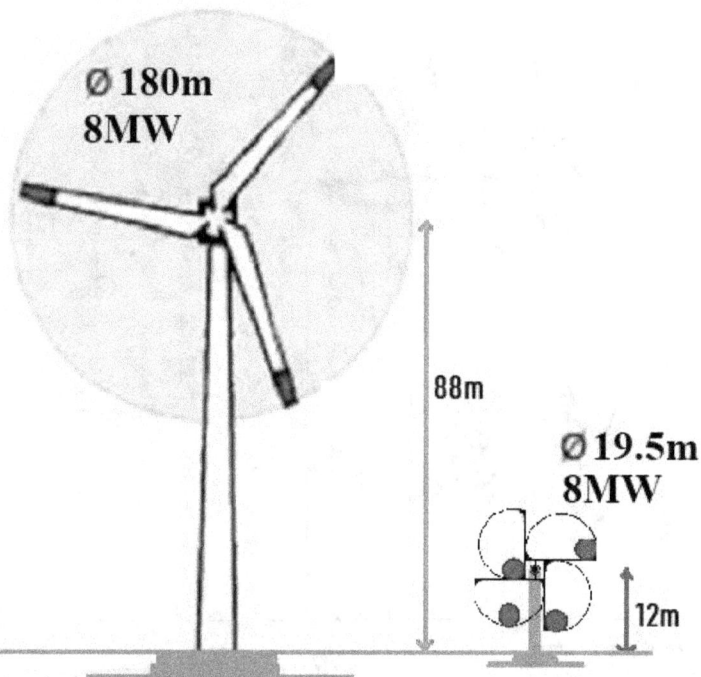

8MW wind turbine
Rotor diameter: 180m
Average rotation speed: 8.5rpm
Weight of rotor and nacelle: 550 tonnes
Mast mass: 550 tonnes
Mast size: 88m
Total mass: 1000 tonnes of steel plus 105 tons of composite materials
Average rate: from 14% to 40% of the time

8MW gravitational turbine
Rotor diameter: 19.73m
Average rotation speed: 5rpm
Weight of rotor and nacelle: 2200 tonnes
Mast mass: 250 tonnes
Mast size: 12m
Total mass: 2274 tons of steel plus 176 tonnes of carbon fiber
Average rate: 98% of the time

3.4 Methodology for sizing a gravitational turbine

Sizing an accurate autonomous gravitational turbine is a relatively straightforward process if you determine all of the parameters to consider.

The design of a gravitational turbine is the result of an optimization carried out by iteration.

The sizing is done through the following steps:

1. Determination of needs:

 Power, stroke $2l$ of the rail, the radius $R_M = 2l / 5$ of the rolling weight, radius $R = R_M$ of the squared circle formed by the four rails, speed of rotation of the turbine, mechanical torque of the turbine, net moment applied to the turbine, the multiplication ratio G of the multiplier used.

2. Determination of the main parameters:

* Driving stroke: $2l$

* Distance k:

$$k^2 = (l + 2R_M)^2 + (l)^2$$

* distance **c:**

$$c^2 = 4R_M^2 + 4l^2$$

* The center of gravity of the rail: : $y_G = \dfrac{2(l+R_M)+\pi R_M}{2\pi}$

* Distance a between the center of gravity of one of the rails and the axis of rotation $a^2 = (y_G + R)^2 + (l)^2$

*Angle**β** :

$$\cos \lambda = \dfrac{k^2+(l+2R_M)^2-(l)^2}{2k(l+2R_M)} \qquad \Rightarrow \lambda = x \deg$$ (According to al-Kashi's Law of Cosines)

$\beta = 180- \lambda \Rightarrow \sin\beta = y$

3. Finding the mass M of the rolling cylindrical weight:

$$M= \dfrac{\Gamma_{TUR\,B}}{g.(2l-K\sin\beta)} \quad \text{avec } M_1=M_2=M_3=M_4=M$$

L =? b =?

4. Inertia of the 4 rolling weights around the axis of the turbine:

$J_{\text{tot-rolling-weights}} = 2\ J_{M1} + 2\ J_{M4} = 2M(c^2+R^2)$

5. Mass of the four rails:

The mass of a rail is given by the following formula after choosing the material and the value of the distance b

$$m_{\text{RAIL}} = \rho bh(2l + \pi l + 2\pi R_M) \quad \text{avec } h = \sqrt{\frac{6Mfz}{b\,\sigma_{maxi}}}$$

$$Mfz = Mg\,\pi\,(l + R_M) \quad \text{with} \quad l \text{ and } R_M \text{ in (mm)}$$

$$\text{And } \sigma_{maxi} = \frac{Re}{S \times S_{fat} \times k_c}$$

With   Re: elastic limit of the used material

$S = 1.5 \Rightarrow$ safety coefficient

$S_{fat} = 1.5 \Rightarrow$ safety coefficient with respect to fatigue

$K_c = 2 \Rightarrow$ shock coefficient

6. Inertia of the four rotating rails around the axis of the turbine:

$J_{\text{Total-rails}} = 4\ J_{\text{Rail}} = 4\ m_{\text{Rail}}\ a^2$

7. Total inertia of the turbine:

$$J_{TURB} = 4\, m_{Rail}\, a^2 + 2M(c^2 + R^2)$$

8. Calculation of the angular and linear acceleration of the turbine:

$$\alpha = \frac{\Gamma_{TURB}}{J_{TURB}} \quad \text{and} \quad a_{Turb} = c\, \alpha$$

With α : angular acceleration of the turbine

a_{Turb} : linear acceleration of the turbine

9. Checking the dynamic equilibrium of rotation of the turbine:

1st condition:

$a_{TURB} < \tilde{a}$ with $\tilde{a}$: the linear acceleration average's a_M of the rolling weight $\tilde{a} = \frac{a_i + a_f}{2}$

2nd condition:

$\Delta t < 0.7 \Delta t_1$

With $\Delta t = \frac{W_{TURB}}{\alpha}$ and $\Delta t_1 = \frac{\pi}{2W_{TURB}} \quad \Rightarrow \quad \frac{\pi}{2W_{TURB}} > \frac{W_{TURB}}{\alpha}$

Δt: the time after which the turbine reaches its speed W_{TURB} thanks to the angular acceleration α of the turbine

Δt₁: the time necessary for the turbine to make the ¼ of a driving revolution

10. Calculation of the maximum admissible inertia of the turbine to secure the system and solve the vibration problem:

* Minimum admissible angular acceleration

$$\alpha_{\text{min-dmissible}} = \frac{W_{TURB}}{\Delta t_{\text{max-admissible}}}$$

With $\Delta t_{\text{max-admissible}} = 0.7 \Delta t_1$

Or directly $\alpha_{\text{min-dmissible}} = \frac{2 W_{TURB}^2}{0.7 \pi}$ since $\Delta t_1 = \frac{\pi}{2 W_{TURB}}$

Note: For a calculation with 2 digits after the decimal point I advise you to use the first formula.

$$J_{MAX-TURB} = \frac{\Gamma_{TURB}}{\alpha_{min-dmissible}} = J_{TURB} + J_A$$

With J_A : added inertia

$$J_A = 4 \times J_{\text{new-rail}}$$

$$m_{\text{new-rail}} = \frac{J_{new-rail}}{a^2} \quad \text{with} : a^2 = (y_G + R)^2 + (l)^2$$

a: Distance between the center of gravity of the main rail and the axis of rotation of the turbine.

4. The new theory of energy production exploiting constant natural forces

In my opinion: Free energy is the energy produced by a system capable of repetitively stimulating a constant natural force and causing it to move periodically and continuously.

The amount of energy required for starting or operating (as the case may be) the technical solution responsible for stimulating the constant natural force is not the only amount of input energy of the system but also the amount of energy produced by the stimulated natural force is an input energy. The output energy is the sum of the two quantities.

The amount of energy is always the same regardless of the physical transformations that take place in the system.

5. Conclusion

I would like to note my objection to the combination of the two words "motion" which denotes the transformation and dissipation of energy and "perpetual" which denotes the violation of the laws of thermodynamics.

An object cannot move if one of these fundamental physical principles is not respected. Motion can only exist if the first and second laws of thermodynamics are respected.

Therefore, the continued adoption of the term "perpetual motion" with its glaring contradiction, leads us to a big scientific error when defining the mechanisms that produce free energy as theoretical systems producing output energy higher than the input energy. This currently adopted analysis is completely wrong because free energy systems generate more energy than they need to function and less than they consume at the input. The meaning is completely different between the two definitions. Free energy is produced by mechanisms that generate more energy than is needed to make it work, and this does not violate the first principle of thermodynamics, since the amount of energy required for the mechanism is only the small amount which satisfies the need to start or operate the technical solution which stimulates the natural constant force surrounding the system and makes it operable as the gravitational force or the force of atmospheric pressure or the force of the hydrostatic pressure of the water, so that the energy entering the system is equal to the output energy and the amount of energy is always the same, no matter what physical transformations take place in this system, this is exactly the first principle of thermodynamics, and during the conversion of energy, a part is converted into heat which increases the overall entropy of the system and this is exactly the second principle of thermodynamics.

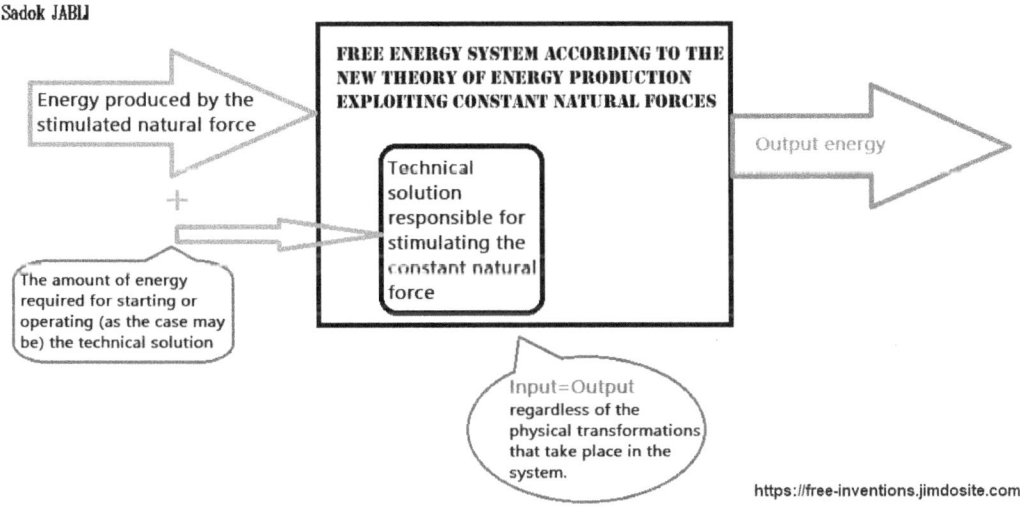